United States Nuclear Regulatory Commission

Protecting People and the Environment

NUREG-1946

Inservice Testing of Pumps and Valves, and Inservice Examination and Testing of Dynamic Restraints (Snubbers) at Nuclear Power Plants

Draft Report for Comment

Manuscript Completed: July 2010
Date Published: July 2010

Prepared by:
S.G. Tingen and G.S. Bedi

ABSTRACT

In this NUREG, the staff of the U.S. Nuclear Regulatory Commission (NRC) discusses the applicable regulations for the inservice testing of pumps and valves and the examination and testing of dynamic restraints (snubbers) at commercial nuclear power plants. The information in NUREG-1482, "Guidelines for Inservice Testing at Nuclear Power Plants," Revision 0, issued April 1995, and Revision 1, issued January 2005, has described this topic in the past. This NUREG report replaces Revision 0 and Revision 1 to NUREG-1482 and is applicable, unless stated otherwise, to all editions and addenda to the American Society of Mechanical Engineers *Code for Operation and Maintenance of Nuclear Power Plants* (OM Code), which Title 10 of the *Code of Federal Regulations* (10 CFR) 50.55a(b) incorporates by reference. In addition, the staff discusses other inservice test program topics such as the NRC process for the review of the OM Code, conditions on the use of the OM Code, and interpretations of the OM Code.

PAPERWORK REDUCTION ACT STATEMENT

The information collections contained in this final regulatory guidance are covered by the requirements of 10 CFR Part 50, "Domestic Licensing of Production and Utilization Facilities," which the Office of Management and Budget (OMB) has approved under OMB control number 3150-0011.

Public Protection Notification

The NRC may neither conduct nor sponsor, and a person is not required to respond to, a request for information or an information collection requirement unless the requesting document displays a currently valid OMB control number.

CONTENTS

PREFACE

NUREG publications consist of reports or brochures on regulatory decisions, results of research, results of incident investigations, and other technical and administrative information. Some of the information herein is similar in appearance to U.S. Nuclear Regulatory Commission staff positions given in a regulatory guide because certain recommendations indicate acceptable alternatives to Code requirements. However, this information is not equivalent to staff positions in a regulatory guide or other generic correspondence because this information is strictly intended for voluntary implementation by licensees.

ABBREVIATIONS

ADAMS	Agencywide Documents Access and Management System
ANSI	American National Standards Institute
ASME	American Society of Mechanical Engineers
B&PV	boiler and pressure vessel
BWR	boiling-water reactor
CFR	*Code of Federal Regulations*
ECCS	emergency core cooling system
FR	*Federal Register*
GL	generic letter
ISI	inservice inspection
IST	inservice testing or inservice test
MOV	motor-operated valve
NEI	Nuclear Energy Institute
NRC	U.S. Nuclear Regulatory Commission
OM Code	Code for Operations and Maintenance of Nuclear Power Plants
PdM	predictive maintenance
RG	regulatory guide
RIS	regulatory issue summary
S/RV	safety/relief valve
TRM	technical requirements manual
TS	technical specification

1. INTRODUCTION

The staff of the U.S. Nuclear Regulatory Commission (NRC) is issuing this NUREG to assist industry in establishing a basic understanding of the regulatory basis for pump and valve inservice testing (IST) programs and dynamic restraint (snubbers) examination and testing programs. This NUREG also provides information regarding the NRC's involvement in the development of the American Society of Mechanical Engineers (ASME) *Code for Operation and Maintenance of Nuclear Power Plants* (OM Code). In this NUREG, the staff discusses OM Code inquiries, the inservice examination and testing of snubbers, pump and valve IST, the use of ASME code cases, conditions on the use of the OM Code, guidance for OM Code noncompliance, and requests for alternatives to the OM Code at operating commercial nuclear power plants.

The information in NUREG-1482, "Guidelines for Inservice Testing at Nuclear Power Plants," Revision 0, issued April 1995, and Revision 1, issued January 2005, has described this topic in the past. This NUREG replaces Revision 0 and Revision 1 to NUREG-1482 and is applicable, unless stated otherwise, to all editions and addenda to the OM Code, which Title 10 of the *Code of Federal Regulations* (10 CFR) 50.55a(b) incorporates by reference.

2. REGULATORY BASIS

The following regulations are applicable to pump and valve IST programs and snubber inservice examination and testing programs at operating reactors and are discussed throughout this NUREG. As a result of the unique wording in various paragraphs, note that the NRC **authorizes** licensee-proposed alternatives in accordance with 10 CFR 50.55a(a)(3), **grants** relief and **imposes** alternative requirements in accordance with 10 CFR 50.55a(f)(6)(i) and 10 CFR 50.55a(g)(6)(i), or **approves** the use of later code editions and addenda in accordance with 10 CFR 50.55a(f)(4)(iv) and 10 CFR 50.55a(g)(4)(iv).

2.1 Incorporation by Reference

The regulations at 10 CFR 50.55a, "Codes and Standards," define the requirements for applying industry codes and standards to boiling- or pressurized-water-cooled nuclear power facilities. The National Technology Transfer and Advancement Act of 1995 (P.L. 104-113) requires that if agencies establish technical standards, they must use technical standards that voluntary consensus standards bodies develop or adopt unless the use of such standards is inconsistent with applicable law or is otherwise impractical. P.L. 104-113 requires Federal agencies to use industry consensus standards to the extent practical; however, it does not require Federal agencies to endorse a standard in its entirety. The law does not prohibit an agency from generally adopting a voluntary consensus standard while taking exception to specific portions of the standard if those provisions are deemed to be "inconsistent with applicable law or otherwise impractical." Furthermore, taking specific exceptions furthers the congressional intent of Federal reliance on voluntary consensus standards because it allows the adoption of substantial portions of consensus standards without the need to reject the standards in their entirety because of limited provisions that are not acceptable to the agency.

The OM Code is a national, voluntary consensus standard, and P.L. 104-113 requires Government agencies to use it unless the use of such a standard is inconsistent with applicable law or is otherwise impractical. The NRC approves or mandates or both the use of editions and addenda to the codes in 10 CFR 50.55a through the rulemaking process of "incorporation by reference." As such, each provision of the codes that 10 CFR 50.55a incorporates by reference and mandates constitutes a legally binding NRC requirement imposed by rule.

The requirements of the OM Code become NRC regulations once they are incorporated by reference into 10 CFR 50.55a(b).

2.2 OM Code Cases

The regulation at 10 CFR 50.55a(b)(6) incorporates by reference Regulatory Guide (RG) 1.192, "Operation and Maintenance Code Case Acceptability, ASME OM Code." Licensees may implement the code cases listed in RG 1.192 without obtaining further NRC review or approval if the code cases are used in their entirety with any supplemental conditions specified in the RG. RG 1.193, "ASME Code Cases Not Approved for Use," lists code cases not approved for use. The NRC staff may authorize licensees to use a proposed alternative to an NRC-approved code case listed in RG 1.192 if requested under 10 CFR 50.55a(a)(3).

The NRC may authorize the use of a code case that it has not yet approved for use in RG 1.192 if a licensee requests the use of the code case under 10 CFR 50.55a(a)(3). The NRC may authorize the use of such a code case until a future revision to RG 1.192 publishes the ASME code case. At that time, if the licensee intends to continue implementing the code case, it must follow all the provisions of the code case with the conditions specified in RG 1.192, if any. The authorization for a specific licensee to use a code case that is not listed in RG 1.192 does not authorize any other licensee to use the code case without it submitting an alternative under 10 CFR 50.55a(a)(3).

2.3 Inservice Testing of Pumps and Valves Program Scope

The regulation at 10 CFR 50.55a(f)(4) requires that throughout the service life of a boiling or pressurized water-cooled nuclear power plant facility, pumps and valves which are classified as ASME Code Class 1, 2, and 3 must meet the inservice test requirements of the ASME OM Code as incorporated by reference in 10 CFR 50.55a(b).

2.4 Inservice Examination and Testing of Snubbers Program Scope

The regulations at 10 CFR 50.55a(g)(4) requires that ASME Code Class 1, 2, and 3 components (including supports) meet the ISI and testing requirements of the Section XI or OM Code as incorporated by reference in 10 CFR 50.55a(b). Article IWF-5000, "Inservice Inspection Requirements for Snubbers," of Section XI of the ASME B&PV Code in the 2005 Addendum and earlier editions and addenda to Section XI provides the requirements for the examination and testing of snubbers in nuclear power plants. Article IWF-5000 was deleted in the 2006 Addendum. Therefore, 10 CFR 50.55a requires inservice examination and testing of snubbers because it incorporates by reference the Section XI requirements in Article IWF-5000 and the OM Code requirements in Subsection ISTD. The inservice examination and testing of snubbers has been a requirement in Article IWF-5000 since Article IWF was first issued in the Winter 1978 Addendum to Section XI. Subsection ISTD of the OM Code has included provisions for the examination and testing of snubbers since it was first issued in 1990.

2.5 Mandatory Inservice Testing and Inservice Inspection Program Updates

The regulation at 10 CFR 50.55a(f)(4)(ii) requires licensees to revise their IST programs every 120 months to reflect the latest edition and addendum to the OM Code incorporated by reference into 10 CFR 50.55a(b)(3) that is in effect 12 months before the start of the new 120-month IST interval.

The regulation at 10 CFR 50.55a(g)(4)(ii) requires licensees to revise their inservice inspection (ISI) programs every 120 months to reflect the latest edition and addendum to Section XI, "Rules for Inservice Inspections of Nuclear Power Plant Components," of the ASME Boiler and Pressure Vessel (B&PV) Code incorporated by reference into 10 CFR 50.55a(b)(2) that is in effect 12 months before the start of the new 120-month ISI interval.

2.6 Alternative Requests

Licensees can request that the NRC authorize an alternative to an OM or Section XI Code requirement in accordance with 10 CFR 50.55a(a)(3). Requests made under 10 CFR 50.55a(a)(3) are more specifically called "alternatives."

2.7 Relief Requests

Licensees can request that the NRC grant relief from an OM or Section XI Code requirement in accordance with 10 CFR 50.55a(f)(5)(iii), 10 CFR 50.55a(f)(5)(iv), 10 CFR 50.55a(g)(5)(iii), and 10 CFR 50.55a(g)(5)(iv). Requests made under 10 CFR 50.55a(f)(5) and 10 CFR 50.55a(g)(5) are called "relief."

2.8 Voluntary Use of Later Editions and Addenda to the ASME Code

The regulation at 10 CFR 50.55a(f)(4)(iv) states that ISTs for pumps and valves may meet the requirements set forth in subsequent editions and addenda to the OM Code that 10 CFR 50.55a(b)(3) incorporates by reference, subject to NRC approval. The regulation at 10 CFR 50.55a(g)(4)(iv) states that the inservice examination of components may meet the requirements set forth in subsequent editions and addenda of the OM or Section XI Code that 10 CFR 50.55a(b) incorporates by reference, subject to NRC approval. This includes the examination and testing of snubbers. Licensees may use portions of editions or addenda provided that all related requirements of the respective editions or addenda are met. When requesting to use editions and addenda to the OM or Section XI Code that have not yet been incorporated by reference, licensees must request authorization to use these later editions and addenda as an alternative to the regulations under 10 CFR 50.55a(a)(3).

2.9 Quality Assurance

IST and ISI programs, including implementing procedures, are subject to the requirements of Appendix B, "Quality Assurance Criteria for Nuclear Power Plants and Fuel Reprocessing Plants," to 10 CFR Part 50, "Domestic Licensing of Production and Utilization Facilities."

2.10 Exemptions

Under 10 CFR 50.12(a), the NRC may, either on its own initiative or upon application by any licensee, grant an exemption from the requirements of 10 CFR Part 50 that is authorized by law, does not present an undue risk to the public health and safety, is consistent with the common defense and security, and is appropriate because of special circumstances. If the NRC approves the application, the exemption relieves the licensee from compliance with the regulation(s) involved. Exemptions are normally not used for 10 CFR 50.55a requests.

3. NRC REVIEW OF THE ASME OM CODE

The first edition and addendum to the OM Code that 10 CFR 50.55a incorporated by reference were the 1995 Edition and the 1996 Addendum. The NRC determines acceptability of new provisions in new editions and addenda to the OM Code and the need for conditions on the use of the OM Code. Generally, the NRC staff participates with other ASME committee members in discussions and technical debates in the development of OM Code revisions. NRC committee representatives discuss the codes and technical justifications with other cognizant NRC staff to ensure an adequate technical review. Finally, NRC management reviews and approves the proposed agency position on the OM Code as part of the rulemaking amending 10 CFR 50.55a to incorporate by reference new editions and addenda to the OM Code and conditions on its use. This process, when considered along with ASME's own process for developing and approving the OM Code, provides reasonable assurance that the NRC approves for use, only those new and revised OM Code editions and addenda (with conditions as necessary), that provide reasonable assurance of adequate protection to public health and safety and that do not have significant adverse impacts on the environment.

4. CONDITIONS ON USE OF THE ASME OM CODE

The OM Code is technically adequate, consistent with current NRC regulations, and approved for use in 10 CFR 50.55a(b)(3), subject to the five conditions outlined below.

4.1 10 CFR 50.55a(b)(3)(i)—Quality Assurance

The OM Code references the use of either the American National Standards Institute (ANSI)/ASME NQA-1-1979, "Quality Assurance Program Requirements for Nuclear Facilities," issued 1979, or the owner's Appendix B to 10 CFR Part 50 quality assurance program as part of its individual provisions for a quality assurance program. However, ANSI/ASME NQA-1-1979 does not contain some of the quality assurance provisions and administrative controls governing operational phase activities that would be required in order to use ANSI/ASME NQA-1-1979 in lieu of an owner's Appendix B quality assurance program description. The NRC originally endorsed ANSI/ASME NQA-1-1979 with the knowledge that it was not entirely adequate and that other commitments such as the ANSI standards must supplement it. Hence, ANSI/ASME NQA-1-1979 is not acceptable for use without the other quality assurance program provisions identified in technical specification (TS) and licensee quality assurance programs.

4.2 10 CFR 50.55a(b)(3)(ii)—Motor-Operated Valve Testing

This condition requires that licensees establish a program to ensure that motor-operated valves (MOVs) continue to be capable of performing their design-basis safety functions. The condition in 10 CFR 50.55a(b)(3)(ii) supplements the stroke-time testing requirement in the 1995 Edition through the 2006 Addendum to Subsection ISTC of the OM Code applicable for MOVs with programs that licensees have previously committed to that demonstrate the design-basis capability of MOVs. Since 1989, the NRC has recognized that the quarterly stroke-time testing requirements for MOVs are not sufficient to provide assurance of MOV operability under design-basis conditions. For example, in Generic Letter (GL) 89-10, "Safety-Related Motor-Operated Valve Testing and Surveillance," dated June 28, 1989, the NRC stated that stroke-time testing alone is not sufficient to provide assurance of MOV operability under design-basis conditions. Therefore, in GL 89-10, the NRC staff requested licensees to verify the design-basis capability of their safety-related MOVs and to establish long-term MOV programs. The NRC subsequently issued GL 96-05, "Periodic Verification of Design-Basis Capability of Safety-Related Power-Operated Valves," dated September 18, 1996, to provide updated guidance for establishing long-term MOV programs.

The provisions in Code Case OMN-1, "Alternative Rules for Preservice and Inservice Testing of Certain Electric Motor-Operated Valve Assemblies in Light-Water Reactor Power Plants," allow users to replace quarterly MOV stroke-time testing with a combination of MOV exercising at least every refueling outage and MOV diagnostic testing on a longer interval. The NRC has determined that Code Case OMN-1 is acceptable for MOVs in lieu of Subsection ISTC of the OM Code and meets the requirements of 10 CFR 50.55a(b)(3)(ii).

The requirements in 10 CFR 50.55a(b)(3)(ii) do not alter expectations on existing licensee commitments relating to MOV design-basis capability. Without being overly prescriptive, this regulation allows licensees to implement the regulatory requirements in a manner that best suits their particular application. This regulation does not require licensees to implement the Joint Owners Group program on MOV periodic verification.

4.3 10 CFR 50.55a(b)(3)(iv)(A), (B), (C), and (D)—Appendix II

This condition supplements the provisions in Appendix II, "Check Valve Condition Monitoring Program," to the OM Code. Subsection ISTC of the OM Code permits the use of Appendix II as an alternative to other testing or examination provisions of Subsection ISTC. If a licensee elects to use Appendix II, the provisions of Appendix II become mandatory in accordance with OM Code requirements. The conditions in 10 CFR 50.55a(b)(3)(iv) do not apply to the 2003 Addendum and later editions and addenda to the OM Code because the 2003 Addendum revised the earlier OM Code provisions on which this regulation was based to address the underlying issues that led the NRC to impose the condition.

The condition in 10 CFR 50.55a(b)(3)(iv)(A) applies to the testing or examination of the check valve obturator movement to both the open and closed positions to assess its condition and confirm acceptable valve performance. The OM main committee approved the bidirectional testing of check valves for inclusion in the 1996 Addendum to the OM Code. The NRC agrees with the need for a required demonstration of the bidirectional exercising movement of the check valve disc. The single direction flow testing of check valves will not always detect degradation of the valve. The classic example of this faulty testing strategy is that the departure of the disc would not be detected during forward flow tests. The departed disc could be lying in the valve bottom or another part of the system and could move to block flow or disable another valve. Appendix II did not require bidirectional testing of check valves in the 1996 through 2002 Addenda to the OM Code. Hence, the condition in 10 CFR 50.55a(b)(3)(iv)(A) was included so that an Appendix II condition monitoring program includes bidirectional testing of check valves to assess their condition and confirm acceptable valve performance (as is required by the OM Code).

The condition in 10 CFR 50.55a(b)(3)(iv)(B) applies to the length of the check valve test interval. Appendix II would permit a licensee to extend check valve test intervals without limit. A policy of prudent and safe interval extension dictates that any interval extension must be based on sufficient experience to justify the additional time. Condition monitoring and current experience may qualify some valves for an initial extension, whereas the trending and evaluation of the data may dictate reduction in the testing interval for some valves. Extensions of IST intervals must consider plant safety and be supported by the trending and evaluation of both generic and plant-specific performance data to ensure that the component is capable of performing its intended function over the entire IST interval. Thus, the condition in 10 CFR 50.55a(b)(3)(iv)(B) limits the time between the initial test or examination and the second test or examination to two fuel cycles or 3 years (whichever is longer), with additional extensions limited to one fuel cycle. The total interval is limited to a maximum of 10 years. An extension or reduction in the interval between tests or examinations would have to be supported by trending and evaluation of performance data.

The condition in 10 CFR 50.55a(b)(3)(iv)(C) applies to a licensee who discontinues a condition monitoring program when using the 1995 Edition of the OM Code with the 1996 and 1997 Addenda. A licensee who discontinues the use of Appendix II is required to implement the requirements of Subsections ISTC 4.5.1 through ISTC 4.5.4 of the OM Code.

The condition in 10 CFR 50.55a(b)(3)(iv)(D) applies to a licensee who discontinues a condition monitoring program when using the 1998 Edition through the 2002 Addendum to the OM Code. A licensee who discontinues the use of Appendix II is required to implement the applicable provisions in Subsection ISTC.

4.4 10 CFR 50.55a(b)(3)(v)—Subsection ISTD

This condition provides requirements for the examination and testing of snubbers. The condition in 10 CFR 50.55a(b)(3)(v) allows licensees using editions and addenda up to the 2005 Addendum of the ASME B&PV Code Section XI, to optionally use Subsection ISTD of the OM Code in place of the requirements for snubbers in Section XI. This condition also states that snubber preservice and inservice examinations must be performed using the VT-3 visual examination method when using Subsection ISTD of the OM Code. The NRC imposed the VT-3 visual examination requirement to ensure that licensees use an appropriate visual examination method for the inspection of integral and nonintegral snubber attachments, such as lugs, bolting, and clamps, when using Subsection ISTD.

Licensees that use the 2006 Addendum and later editions and addenda to Section XI of the ASME B&PV Code must follow the requirements of Subsection ISTD of the OM Code for snubbers because ASME removed the requirements for the examination of snubbers from the scope of Section XI in the 2006 Addendum. The condition in 10 CFR 50.55a(b)(3)(v) does not invoke the VT-3 visual examination requirement when licensees use the 2006 Addendum and later editions and addenda to Section XI because ASME revised Figure IWF-1300-1 in the 2006 Addendum to Section XI to clarify that integral and nonintegral snubber attachments are within the scope of Section XI. Therefore, the visual examination method specified in the 2006 Addendum and in later editions and addenda to Section XI applies to the examination of integral and nonintegral snubber attachments.

4.5 10 CFR 50.55a(b)(3)(vi)—Exercise Interval for Manual Valves

This condition requires that manual valves must be exercised on a 2-year interval rather than the 5-year interval specified in paragraph ISTC-3540 of the 1999 through 2005 Addenda to the OM Code, provided that adverse conditions do not require more frequent testing. The 1998 Edition and earlier versions of the OM Code specified an exercise interval of 3 months for manual valves. The 1999 Addendum to the OM Code revised paragraph ISTC-3540 to extend the exercise frequency for manual valves to 5 years; however, the NRC staff did not agree that there was sufficient justification to extend the exercise interval for manual valves to 5 years. The condition in 10 CFR 50.55a(b)(3)(vi) does not apply to the 2006 Addendum to the OM Code because ASME revised the exercise interval in paragraph ISTC-3540 of the 2006 Addendum to the OM Code to 2 years for manually-operated valves.

5. DEVELOPING AND IMPLEMENTING 10 CFR 50.55a PROGRAMS

5.1 Alternative Requests

The OM Code establishes the requirements for preservice testing and IST and the examination of certain components to assess their operational readiness in light-water reactor nuclear power plants. These requirements apply to safety-related pumps, valves, pressure relief devices, and snubbers. The requirements are constantly being reviewed and improved in order to meet the basic function of maintaining the safe and reliable operation and maintenance of nuclear power plants.

It is understood that not all plants are designed the same. It is also understood that the general requirements developed in the OM Code may not be applicable or that complying with these requirements may be difficult. Licensees may use proposed alternatives to the OM Code provided that (1) the alternative would provide an acceptable level of quality and safety under 10 CFR 50.55a(a)(3)(i); or (2) compliance with the specified requirements would result in hardship or unusual difficulty without a compensating increase in the level of quality and safety under 10 CFR 50.55a(a)(3)(ii). Hardships generally involve reductions in radiation exposure to as low as reasonably achievable, challenges to operators or plant equipment, components that are somewhat unique in design such as jockey pumps, or systems where pump flow is fixed and cannot be adjusted.

Licensees cannot implement proposed alternatives to the OM Code requirements under 10 CFR 50.55a(a)(3) until the NRC staff completes its evaluation. For example, if a licensee proposes to implement a pump vibration program based on the use of spectral analysis rather than the OM-Code-specified method, the licensee must continue to meet the OM Code requirements until the NRC staff completes its evaluation.

5.2 Relief Requests

The regulation at 10 CFR 50.55a(f)(4) requires licensees to test pumps and valves in the IST program to the "extent practical" within the limitations of the design, geometry, and materials of construction. The regulations at 10 CFR 5.55a(f)(5)(iii)–(iv) and 10 CFR 50.55a(f)(6)(i) use the term "impractical" instead of "extent practical." The terms "extent practical" and "impractical" apply to test requirements in the OM Code that licensees cannot perform due to the design, geometry, and materials of construction of the pump or valve. For example, the new comprehensive pump test provisions in Subsection ISTB of the 1994 Addendum to the OM Code require licensees to test pumps in the IST program at design flow conditions. These new test provisions are in the 1995 Edition of the OM Code, which 10 CFR 50.55a incorporates by reference. In some instances, facilities were not designed to test pumps at design flow conditions. Some licensees may have difficulty fully implementing these tests, and, in certain cases, because of the impracticality of implementation, a request for relief under 10 CFR 50.55a(f)(5) would be appropriate.

In accordance with the regulations, when updating a program to a later edition of the OM Code, licensees must implement the updated program at the beginning of a 120-month interval. The regulations state that in cases in which a licensee determines that an OM-Code-specified pump or valve test is impractical and is not included in the revised IST program, it must submit a relief request demonstrating the basis for its determination to the NRC no later than 12 months after the 120-month interval ends. However, experience has shown that licensees also identify impractical test provisions throughout the interval. In such cases, licensees may request relief as soon as they identify the condition. Because the requirements are impractical, the licensee would test the applicable components using the method proposed in the relief request in the period of time from the beginning of the new interval (or from the time of identification) until the NRC staff completes its evaluation.

5.3 Voluntary Use of Later Editions of and Addenda to the ASME OM Code

Licensees must conduct ISTs during successive 120-month intervals to verify the operational readiness of pumps and valves within the scope of the OM Code in accordance with 10 CFR 50.55a(f)(4)(ii). In conducting these ISTs, licensees must comply with the provisions of the latest edition and addendum to the OM Code, which 10 CFR 50.55a(b) incorporates by reference, 12 months before the start of the 120-month interval, subject to the limitations and modifications listed in 10 CFR 50.55a(b).

The regulation at 10 CFR 50.55a(f)(4)(iv) states that ISTs for pumps and valves may meet the requirements set forth in subsequent editions and addenda to the OM Code that 10 CFR 50.55a(b)(3) incorporates by reference, subject to NRC approval. The regulation at 10 CFR 50.55a(g)(4)(iv) states that the inservice examination of components may meet the requirements set forth in subsequent editions and addenda of the OM or Section XI Code that 10 CFR 50.55a(b) incorporates by reference, subject to NRC approval. This includes the examination and testing of snubbers. Licensees may use portions of editions or addenda provided that all related requirements of the respective editions or addenda are met. When requesting to use editions and addenda to the OM or Section XI Code that have not yet been incorporated by reference, licensees must request authorization to use these later editions and addenda as an alternative to the regulations under 10 CFR 50.55a(a)(3).

The amount of written documentation needed for a request to use a later OM Code edition and addendum that 10 CFR 50.55a(b) incorporates by reference is significantly less than that of a request to use an alternative requirement. For example, licensees are not required to justify requests to use the later OM Code editions and addenda that 10 CFR 50.55a(b) incorporates by reference. In contrast, when submitting an alternative request, licensees must provide justification that the proposed alternative would provide an acceptable level of quality and safety. If a licensee uses portions of a later OM Code edition and addendum, it must ensure that all related requirements of the respective editions and addenda are met. The licensee should discuss the related requirements in its letter to the NRC. The regulations do not specify when the licensee should submit the letter, only that it should submit the letter before it uses the later OM Code edition and addendum. The staff issued Regulatory Issue Summary (RIS) 2004-12, "Clarification on Use of Later Editions and Addenda to the ASME OM Code and Section XI," dated July 28, 2004, in order to clarify this matter.

5.4 Identification of Code Noncompliance

RIS 2005-20, Revision 1, "Revision to NRC Inspection Manual Part 9900 Technical Guidance, Operability Determinations & Functionality Assessments for Resolution of Degraded or Nonconforming Conditions Adverse to Quality or Safety," dated April 16, 2008, provides guidance on resolving degraded and nonconforming conditions for those components in the IST program. Section 6.2 of the NRC Inspection Manual, Part 9900, "Technical Guidance," notes that a licensee may discover a noncompliance with a regulation and the noncompliance should be treated as a degraded or nonconforming condition. The licensee should assess the operability or functionality of the affected component. If the operating license or the TS does not address the noncompliance (i.e., the noncompliance has no impact on any specified safety function), the licensee should determine if the noncompliance raises an immediate safety issue. Immediate action, such as shutting down the plant, may not be required unless otherwise specified by NRC requirements. The licensee should determine if any other NRC requirements apply to the situation (e.g., Criterion XVI, "Corrective Action," of Appendix B to 10 CFR Part 50 or 10 CFR 50.12, "Specific Exemptions") and take any action required. A licensee may submit a request for an alternative under 10 CFR 50.55a(3) for a component declared inoperable due to an OM Code nonconformance, and the NRC will evaluate the alternative. The inoperable component could be declared operable once the NRC authorizes the alternative and the licensee has successfully completed the alternative test (if applicable). An NRC authorization of the alternative would not be retroactive because the agency has to authorize it before it can be implemented.

5.5 Revising NRC-Authorized Alternatives

The NRC must authorize any change to an NRC-authorized 10 CFR 50.55a(a)(3) alternative unless the requirements of the ASME code can be met. For example, many licensees created a technical requirement manual (TRM) to control certain provisions relocated from TS. Licensees relocated snubber examination and testing requirements from the TS to the TRM. The TRM requirements are controlled using the criteria in 10 CFR 50.59, "Changes, Tests and Experiments." The regulations at 10 CFR 50.59 require licensees to evaluate proposed changes to their facilities for the effects of these changes on the licensing basis of the plant, as described in the final safety analysis report (as updated) and to obtain prior NRC approval for changes that meet specified criteria as having a potential impact upon the basis for the issuance of the operating license. In the case of snubber examination and testing, the NRC has authorized the use of the TRM snubber examination and testing requirements in lieu of the ASME code requirements at numerous operating plants. The NRC authorized the use of the requirements contained in the TRM as an alternative to the ASME code requirements. The use of an alternative as authorized by the NRC becomes a regulatory requirement; therefore, the NRC staff must review and approve changes to these requirements under 10 CFR 50.55a(a)(3).

Nuclear Energy Institute (NEI) 96-07, "Guidelines for 10 CFR 50.59 Implementation," Revision 1, dated November 2000, states that licensees' activities that are controlled by the regulations at 10 CFR 50.55a take precedence over 10 CFR 50.59. RG 1.187, "Guidance for Implementation of 10 CFR 50.59, Changes, Tests, and Experiments," issued November 2000, endorses NEI 96-07, Revision 1. Similarly, Section D, "Implementation," of RG 1.187 states that 10 CFR 50.59 cannot be used in those cases in which a licensee proposes an acceptable alternative method for complying with the specified portion of the NRC's regulation.

5.6 NRC Verbal Authorization of an Alternative Request

On rare occasions, the NRC may grant verbal authorizations as an alternative under 10 CFR 50.55a(a)(3) when, because of unforeseen circumstances, licensees need NRC authorization before the agency is able to issue its written safety evaluation. Temporary verbal authorization for an alternative under 10 CFR 50.55a(a)(3) is subject to the following:

- The proposed alternative is in writing, and all the information that the NRC requires to complete the safety evaluation has been docketed.

- An identified need for the verbal authorization is recognized given the circumstances of the licensee's request.

- The NRC has completed its review and determined that the proposed alternative is technically justified, but the agency has not yet formally documented it in a safety evaluation.

- The technical branch and reactor licensing branch chiefs have agreed to the verbal authorization.

Verbal authorization is most likely conveyed in a telephone conversation. As such, appropriate NRC personnel who are normally involved in authorizing the alternative must be present in the telephone conversation. The NRC project manager should promptly (i.e., in 1 or 2 days) generate a record of the conversation; this record will meet the definition of an official agency record and must be entered into the NRC's Agencywide Documents Access and Management System (ADAMS) and made publicly available. The NRC should issue the final written authorization within 150 days after giving verbal authorization.

5.7 Submittal of Previously NRC-Authorized Alternatives

Licensees occasionally submit alternative requests that are very similar to NRC-authorized alternative requests for the previous 10-year IST intervals when updating their IST program in accordance with 10 CFR 50.55a(f)(4)(ii). This practice is acceptable provided that the licensee compares the requirements between the old and new OM Codes and evaluates whether changes to the alternative request are necessary. For example, the OM Code has new provisions added for exercising check valves such as disassembly and condition monitoring programs. Addressing the check valve disassembly and condition monitoring programs in the alternative request may be appropriate if these provisions were not included in the OM Code upon which the original alternative request was based. Furthermore, the addition of disassembly and condition monitoring programs to the OM Code may eliminate the need for the alternative request.

Licensees should also review new code cases before submitting an alternative request for updated IST programs. For example, Code Case OMN-9, "Use of Pump Curve for Testing," provides an alternative method for testing centrifugal and vertical line shaft pumps when the licensee is unable to obtain a specific reference value in accordance with Subsection ISTB of the OM Code. The NRC conditionally approved Code Case OMN-9 in RG 1.192, Revision 0.

Code Case OMN-16, "Use of Pump Curve for Testing," has superseded Code Case ONM-9. The use of Code Cases OMN-9 or OMN-16 may eliminate the need for an alternative request.

5.8 Standard Format for Requests

To improve the effectiveness and efficiency of the request process, NEI developed a white paper entitled, "Standard Format for Requests from Commercial Reactor Licensees Pursuant to 10 CFR 50.55a," Revision 1, dated June 7, 2004 (ADAMS Accession No. ML070100400). The guidance provided to licensees is voluntary.

5.9 ASME OM Code Interpretations

Users of the OM Code may submit technical inquires to the OM main committee. Technical inquiries include requests for revisions to present code requirements or new or additional code requirements, requests for code cases, and requests for code interpretations. Code interpretations provide the meaning or the intent of existing requirements in the OM Code, and after appropriate committee deliberations and balloting, ASME issues responses as clarifications, new or additional code requirements, or a code erratum. Licensees should exercise caution when applying interpretations because they are not specifically incorporated by reference into 10 CFR 50.55a and have not received NRC approval. The NRC recognizes that ASME is the official interpreter of the OM Code, but the NRC will not accept ASME interpretations that, in the NRC's opinion, are contrary to the agency's requirements or may adversely impact facility operations.

On November 12, 1996, representatives from ASME and the NRC met, in part because of the continuing questions from the industry about ASME interpretations. The NRC and ASME discussed how the agency would inform ASME of code interpretations to which it takes exception. It was agreed that the NRC should not establish a formal method for reviewing interpretations for acceptance. It was agreed that any concerns that the NRC has on specific ASME code interpretations would be brought to ASME's attention through the NRC staff's normal interaction with the code. This practice has been routine for many years.

An interpretation that changes the requirements of the OM Code following the incorporation by reference of a particular edition or addendum in 10 CFR 50.55a may not be accepted. An example of an OM Code interpretation that did not conflict with NRC regulatory requirements is Interpretation 01-18, "ASME OM Code-1995 with OMa ASME Code-1996 Addenda, Appendix I," dated June 26, 2003. Subparagraph I-1.1.3(a), "5-Year Test Interval," of Appendix I to the 1995 Edition with the 1996 Addendum to the OM Code requires that Class 1 pressure relief valves be tested at least once every 5 years. Interpretation 01-18 clarifies that the 5-year test interval starts when the valve is tested. Therefore, licensees must test Class 1 pressure relief valves every 5 years (test to test) unless the NRC has authorized a different interval in accordance with 10 CFR 50.55a(a)(3) or unless a different interval is specified in a code case approved for use by the NRC in accordance with 10 CFR 50.55a(b)(6).

5.10 Risk-Informed Inservice Testing

RG 1.175, "An Approach for Plant-Specific, Risk-Informed Decisionmaking: Inservice Testing," issued August 1998, describes an acceptable alternative approach for applying risk insights from probabilistic risk assessment, in conjunction with established traditional engineering information, to make changes to a nuclear power plant's IST program. The approach described in RG 1.175 addresses the high-level safety principles specified in RG 1.174, "An Approach for Using Probabilistic Risk Assessment in Risk-Informed Decisions on Plant-Specific Changes to the Licensing Basis," Revision 1, issued November 2002, and attempts to strike a balance between defining an acceptable process for developing risk-informed IST programs without being overly prescriptive. The resultant risk-informed IST programs will have improved effectiveness with regard to the use of plant resources while still maintaining acceptable levels of quality and safety. However, licensees may propose other approaches for consideration by the NRC staff. The approach presented in RG 1.175 should be regarded as an example of acceptable practices and that licensees should have some degree of flexibility in satisfying regulatory requirements on the basis of their accumulated plant experience and knowledge.

5.11 Use of Code Cases

Code cases are typically formatted in terms of an "Inquiry" (as from a licensee) and "Reply" (by the applicable ASME code committee). Oftentimes, the "Inquiry" and "Reply" are accompanied by an "Applicability" statement that identifies the specific editions and addenda to the OM Code to which the "Reply" applies. However, the OM Code committees do not always (1) include an "Applicability" statement or (2) identify all of the editions and addenda to which the "Reply" applies. As a result, several licensees have questioned whether they could use certain code cases without additional interaction with the NRC.

If a licensee would like to use a code case with an edition or addendum to the OM Code to which it is not applicable, the licensee has the one of the following options:

- The licensee can request the alternative to use the code case beyond its stated applicability and have the NRC authorize the code case under 10 CFR 50.55a(a)(3).

- If the code case is applicable to an edition or addendum of the OM Code that is later than the version of the OM Code being used by the licensee, the licensee could update to the later version of the OM Code under 10 CFR 50.55a(f)(4)(iv) and then use the code case, provided that the code case has been approved for use in RG 1.192. Note that the later version of the OM Code must also have been incorporated by reference into 10 CFR 50.55a, that the licensee must update all related requirements of the respective edition or addendum, and that the NRC must specifically approve the use of the update.

Licensees should not use code cases with editions and addenda to the OM Code to which they do not apply and that are not specifically incorporated by reference in 10 CFR 50.55a(b).

6. PUMP AND VALVE ALTERNATIVE TESTS

6.1 Waterleg Pumps

The NRC has received proposed alternatives from licensees of boiling-water reactors (BWRs) for Group A tests for waterleg pumps. Subsection ISTB-3400 and Table ISTB-3400-1 of the OM Code specify that a Group A test must be performed quarterly for Group A pumps. The waterleg pumps are low flow pumps that are required to operate whenever their respective emergency core cooling system (ECCS) trains are in the operable condition. As such, the pumps perform continuous duty on a recirculation line and provide makeup as needed. There is typically no flow instrumentation of the recirculation line, and the flow instrumentation on the main ECCS header is not accurate enough to measure the low flow of the pumps. When requesting an alternative Group A test for a waterleg pump, a licensee should explain how it monitors pump discharge pressure, how it verifies that the main ECCS header is full of water, and what is the pump vibration monitoring frequency.

6.2 Smooth-Running Pumps

Pumps that have very low vibration reference values (less than or equal to 0.05 inch per second) are called smooth-running pumps. A small increase in smooth-running pump vibration during the OM Code-required IST causes the pump to exceed OM Code vibration acceptance criteria, which normally results in unnecessary corrective action. The NRC has authorized alternative vibration acceptance criteria for smooth-running pumps on a case-by-case basis in accordance with 10 CFR 50.55a(a)(3).

Alternative requests for smooth-running pumps should specify a minimum vibration reference value (0.05 inch per second), combined with including smooth-running pumps in a predictive maintenance (PdM) program. A plant using an NRC-authorized alternative vibration acceptance criteria noted an increasing upward trend in vibration during IST. Corrective action was not required because the vibration was below the alert range. Pump monitoring as part of the plant's PdM program also detected problems but no corrective action was taken. After the pump bearing failed, it became clear that a simple minimum vibration reference value alone is not sufficient to identify degradation of a smooth-running pump. PdM programs normally include bearing temperature trending, oil sampling and analysis, thermographic analysis, and enhanced vibration monitoring. The objective of the PdM program should be to detect and correct problems involving the mechanical condition of the pump before the pump reaches its overall vibration alert limit.

6.3 Vibration-Measuring Transducers

Subsection ISTB of the OM Code requires that the frequency response range of vibration-measuring transducers and their readout system be from one-third of the minimum pump shaft rotational speed to at least 1,000 hertz. Licensees have proposed alternatives to this OM Code requirement in accordance with 10 CFR 50.55a(a)(3) for pumps with low shaft rotational speeds. Similar alternative requests submitted by licensees are now being withdrawn following discussion with the NRC. The proposed alternatives state that the procurement and calibration of vibration-measuring transducers and their readout systems for the lower end of the OM Code-specified range were hardships because of the limited number of vendors supplying such equipment, the level of equipment sophistication, and equipment cost. The NRC typically authorized these alternative requests in the past. However, vibration-measuring transducers and their readout system can now be procured from various suppliers at a reasonably low cost due to technology advancement and research work performed in the field of instrumentation. Therefore, licensee requests to use this alternative are generally no longer authorized by the NRC.

6.4 Online Check Valve Sample Disassembly and Inspection

Licensees have proposed, as an alternative to ISTC-5221(c) and ISTC-5224, to perform sample disassembly and inspection of check valves in a group online. Subsection ISTC of OM Code, Paragraph ISTC-3510 requires that check valves be exercised every 3 months. Paragraph ISTC-3522(c) states that if exercising is not practicable during operation at power and cold shutdown, it shall be performed during refueling outages. ISTC-5221(c) allows disassembly of check valves every refueling outage as an alternative means to verify their operability. Instead of disassembly every refueling outage, ISTC-5221(c) provides the option of using a sample disassembly and inspection program for groups of identical valves in similar applications. Further, ISTC-5221(c)(3) states that at least one valve from each group shall be disassembled and examined at each refueling outage and all valves in each group shall be disassembled and examined at least once every 8 years. ISTC-5224 requires that check valves in a sample disassembly program that are not capable of being full-stroke exercised or have failed or have unacceptably degraded valve internals, shall have the cause of failure analyzed and the condition corrected.

ISTC-5224 also states that other check valves in the sample group that may also be affected by this failure mechanism be examined or tested during the same refueling outage to determine the condition of internal components and their ability to function. A licensee should fully describe how it plans to comply with the requirements in ISTC-5224 when submitting alternative requests for check valve group sample disassembly and inspection online. The plan description also should include information on management of examination and testing of all group valves should a scheduled valve inspection be declared inoperable. For example, licensees should explain how the disassembly and inspection of the other check valves in a group will be completed within the allowed system outage time.

6.5 Safety/Relief Valves

Many licensees have requested and obtained NRC authorization in accordance with 10 CFR 50.55a(a)(3) to use the provisions in Code Case OMN-17, "Alternative Rules for Testing ASME Class 1 Pressure Relief/Safety Valves," as an alternative to the 5-year test interval specified in the OM Code. ASME published Code Case OMN-17 in the 2009 Edition of the OM Code. The NRC plans to include Code Case OMN-17 in RG 1.192 during a future revision. Code Case OMN-17 allows an extension of the test frequency for safety relief valves (S/RVs) from 60 months to 72 months plus a 6-month grace period. The code case imposes a special maintenance requirement to disassemble and inspect each valve to verify that parts are free from defects resulting from the time-related degradation or maintenance-induced wear before the start of the extended test frequency. Although the OM Code does not require that S/RVs be routinely refurbished, refurbishment provides reasonable assurance that the S/RVs are operationally ready during the extended test interval.

In recent years, the NRC staff has received numerous requests for relief or TS changes or both related to the stroke testing requirements for BWR dual-function main steam S/RVs. The 2003 Addendum and earlier editions and addenda to Mandatory Appendix I to the OM Code require the stroke testing of S/RVs after they are reinstalled following maintenance activities. A number of licensees have determined that in situ testing of the S/RVs can contribute to undesirable seat leakage of the valves during subsequent plant operation and have received approval to perform stroke testing at a laboratory facility coupled with in situ tests and other verifications of actuation systems as an alternative to the testing required by the OM Code. The revised subparagraph I-3410(d) in Mandatory Appendix I to the 2004 Edition of the OM Code does not require licensees to stroke test S/RVs at reduced or normal system pressure following maintenance. Subparagraph I-3410(d) in the 2004 Edition of the OM Code requires that each S/RV that has been removed for maintenance or testing and reinstalled shall have the electrical and pneumatic connections verified either through mechanical/electrical inspection or testing before the resumption of electric power generation. Several licensees have requested and obtained NRC approval in accordance with 10 CFR 50.55a(f)(4)(iv) to use Subparagraph I-3410(d) of the 2004 Edition of the OM Code in place of Subparagraph I-3410(d) of the 2001 Edition through the 2003 Addenda to the OM Code.

7. SNUBBER INSERVICE EXAMINATION AND TESTING PROGRAMS

7.1 Program Controls

Some licensees have incorrectly interpreted that the examination and testing of snubbers is not a 10 CFR 50.55a requirement because (1) 10 CFR 50.55a(g) addresses components (including supports) without mentioning snubbers, (2) snubber examination and testing was historically covered by TS, and (3) TS allow snubber examination and testing requirements to be relocated from the TS to the TRM. Licensees have the option to control the inservice examination and testing of snubbers through their TS or other licensee-controlled documents. For plants using their TS to govern the inservice examination and testing of snubbers, 10 CFR 50.55a(g)(5)(ii) requires that if a revised ISI program for a facility conflicts with the TS, the licensee shall apply to the Commission for the amendment of the TS to conform the TS to the revised program. Therefore, when performing 120-month program updates in accordance with 10 CFR 50.55a(g)(4), licensees must submit any required amendments or any alternative requests to ensure that their TS remain consistent with the new ISI program. The TS, TRM, or other licensee-controlled documents governing the snubber inservice examination and testing program do not eliminate the 10 CFR 50.55a requirements to update the program at 120-month intervals in accordance with 10 CFR 50.55a(g)(4) or to request and receive NRC authorization for alternatives to the Code requirements when appropriate.

7.2 OM Part 4 Clarification

The NRC has noted that the relocation of the reference to ASME/ANSI *Operation and Maintenance of Nuclear Power Plants* (OM Part 4) from IWF-5000 of Section XI to Table IWF-1600-1 of Section XI has created confusion regarding which edition and addenda of OM Part 4 must be used. For clarification, the ASME OM Code and OM Part 4 are two different ASME documents. Article IWF-5000 of the 1987 Addendum through the 1992 Edition of Section XI requires that the inservice examination and testing of snubbers be accomplished in accordance with the 1987 Edition of OM Part 4. The reference to OM Part 4 was deleted from IWF-5000 in the 1992 Addendum of Section XI and there is no reference to OM Part 4 in IWF-5000 in the 1992 through 2005 Addenda of Section XI. The reference for the applicable edition and addenda of OM Part 4 was moved to Table 1600-1 in the 1992 Addendum of Section XI. Although IWF-5000 in the 1992 through 2005 Addenda of Section XI no longer references OM Part 4, Table 1600-1 of the 1992 through 2005 Addenda of Section XI requires that inservice examination and testing of snubbers be performed in accordance with the 1987 Edition and 1988 Addendum of OM Part 4.

8. REFERENCES

8.1   American Society of Mechanical Engineers/American National Standards Institute, "Code for Operation and Maintenance of Nuclear Power Plants," New York, NY.

8.2   American Society of Mechanical Engineers/American National Standards Institute, "Operation and Maintenance of Nuclear Power Plants," New York, NY.

8.3   American Society of Mechanical Engineers/American National Standards Institute, NQA-1, "Quality Assurance Program Requirements for Nuclear Facilities," Washington, DC, and New York, NY, 1979.

8.4   American Society of Mechanical Engineers, *Boiler and Pressure Vessel Code*, Section XI, "Rules for Inservice Inspection of Nuclear Power Plant Components," New York, NY.

8.5   Nuclear Energy Institute, NEI-96-07, "Guidelines for 10 CFR 50.59 Implementation," Revision 1, Dated November 2000, Washington, DC.

8.6   Nuclear Energy Institute White Paper, "Standard Format for Requests from Commercial Reactor Licensees Pursuant to 10 CFR 50.55a," Revision 1, Washington, DC, June 7, 2004.

8.7   Part 9900, "Technical Guidance," NRC Inspection Manual

8.8   The National Technology Transfer and Advancement Act of 1995 (P.L. 104-113)

8.9   *U.S. Code of Federal Regulations*, "Domestic Licensing of Production and Utilization Facilities," Part 50, Chapter 1, Title 10, "Energy."

8.10   U.S. Nuclear Regulatory Commission, Generic Letter 89-10, "Safety-Related Motor-Operated Valve Testing and Surveillance," dated June 28, 1989.

8.11   U.S. Nuclear Regulatory Commission, Generic Letter 96-05, "Periodic Verification of Design-Basis Capability of Safety-Related Power-Operated Valves," dated September 18, 1996.

8.12   U.S. Nuclear Regulatory Commission, NUREG-1482, "Guidelines for Inservice Testing at Nuclear Power Plants," Revision 0, Washington, DC, April 1995.

8.13   U.S. Nuclear Regulatory Commission, NUREG-1482, "Guidelines for Inservice Testing at Nuclear Power Plants," Revision 1, Washington, DC, January 2005.

8.14   U.S. Nuclear Regulatory Commission, Regulatory Guide 1.175, "An Approach for Plant-Specific, Risk-Informed Decisionmaking: Inservice Testing," Washington, DC, August 1998.

8.15   U.S. Nuclear Regulatory Commission, Regulatory Guide 1.174, "An Approach for Using Probabilistic Risk Assessment in Risk-Informed Decisions on Plant-Specific Changes to the Licensing Basis," Revision 1, Washington, DC, November 2002.

8.16   U.S. Nuclear Regulatory Commission, Regulatory Guide 1.187, "Guidance for Implementation of 10 CFR 50.59, Changes, Tests, and Experiments," Washington, DC, November 2000.

8.17   U.S. Nuclear Regulatory Commission, Regulatory Guide 1.192, "Operation and Maintenance Code Case Acceptability: ASME OM Code," Washington, DC, June 2003.

8.18   U.S. Nuclear Regulatory Commission, Regulatory Guide 1.193, "ASME Code Cases Not Approved for Use," Washington, DC, June 2003.

8.19   U.S. Nuclear Regulatory Commission, Regulatory Issue Summary 2004-12, "Clarification on Use of Later Editions and Addenda to the ASME OM Code and Section XI," Washington, DC, July 28, 2004.

8.20   U.S. Nuclear Regulatory Commission, Regulatory Issue Summary 2005-20, "Revision to NRC Inspection Manual Part 9900 Technical Guidance, 'Operability Determinations & Functionality Assessments for Resolution of Degraded or Nonconforming Conditions Adverse to Quality or Safety,'" Revision 1, Washington, DC, April 16, 2008.